Dr. Krunalkumar R. Mehariya
Bhagwati K. Gauni Mehariya
Kirti kumar Goswami
Dr. Dipak Vachhani
Sunil Galani
Dr. Ranjit S. Pada

X-Ray Crystallographic Study of Novel Oxazole Derivatives

Anchor Academic Publishing

Mehariya, Krunalkumar R., Mehariya, Bhagwati K. Gauni, Goswami, Kirti kumar, Vachhani, Dipak, Galani, Sunil, Pada, Ranjit S.: X-Ray Crystallographic Study of Novel Oxazole Derivatives, Hamburg, Anchor Academic Publishing 2016

Buch-ISBN: 978-3-96067-083-4
PDF-eBook-ISBN: 978-3-96067-583-9
Druck/Herstellung: Anchor Academic Publishing, Hamburg, 2016

Bibliografische Information der Deutschen Nationalbibliothek:
Die Deutsche Nationalbibliothek verzeichnet diese Publikation in der Deutschen Nationalbibliografie; detaillierte bibliografische Daten sind im Internet über http://dnb.d-nb.de abrufbar.

Bibliographical Information of the German National Library:
The German National Library lists this publication in the German National Bibliography. Detailed bibliographic data can be found at: http://dnb.d-nb.de

All rights reserved. This publication may not be reproduced, stored in a retrieval system or transmitted, in any form or by any means, electronic, mechanical, photocopying, recording or otherwise, without the prior permission of the publishers.

Das Werk einschließlich aller seiner Teile ist urheberrechtlich geschützt. Jede Verwertung außerhalb der Grenzen des Urheberrechtsgesetzes ist ohne Zustimmung des Verlages unzulässig und strafbar. Dies gilt insbesondere für Vervielfältigungen, Übersetzungen, Mikroverfilmungen und die Einspeicherung und Bearbeitung in elektronischen Systemen.

Die Wiedergabe von Gebrauchsnamen, Handelsnamen, Warenbezeichnungen usw. in diesem Werk berechtigt auch ohne besondere Kennzeichnung nicht zu der Annahme, dass solche Namen im Sinne der Warenzeichen- und Markenschutz-Gesetzgebung als frei zu betrachten wären und daher von jedermann benutzt werden dürften.

Die Informationen in diesem Werk wurden mit Sorgfalt erarbeitet. Dennoch können Fehler nicht vollständig ausgeschlossen werden und die Diplomica Verlag GmbH, die Autoren oder Übersetzer übernehmen keine juristische Verantwortung oder irgendeine Haftung für evtl. verbliebene fehlerhafte Angaben und deren Folgen.

Alle Rechte vorbehalten

© Anchor Academic Publishing, Imprint der Diplomica Verlag GmbH
Hermannstal 119k, 22119 Hamburg
http://www.diplomica-verlag.de, Hamburg 2016
Printed in Germany

Contents

1. Introduction 5

2. Oxazole-containing natural products 6

3. Chemistry 8
   3.1 Synthesis of 5-(3-methoxy-4-(prop-2-yn-1 yloxy) phenyl) oxazole 8
   3.2 Preparation of 5-(3-methoxy-4-(prop-2-yn-1 yloxy) phenyl) oxazole (6) .. 8
   3.3 Analytical data 8

4. X-ray Structure Report 9
   4.1 Experimental 9
       4.1.1 Data Collection 9
       4.1.2 Data Reduction 9
       4.1.3 Structure Solution and Refinement 10
   4.2 Experimental Details 10
       4.2.1 Crystal Data 10
       4.2.2 Intensity Measurements 11
       4.2.3 Structure Solution and Refinement 11
       4.2.4 Atomic coordinates and Biso/Beq 12
       4.2.5 Atomic coordinates and Biso involving hydrogen atoms 13
       4.2.6 Anisotropic displacement parameters 13
       4.2.7 Bond lengths (Å) 14
       4.2.8 Bond lengths involving hydrogens (Å) 15
       4.2.9 Bond angles (°) 15
       4.2.10 Bond angles involving hydrogens (°) 16
       4.2.11 Torsion Angles (°) 16
       4.2.12 Intramolecular contacts less than 3.60 Å involving hydrogens 18
       4.2.13 Intermolecular contacts less than 3.60 Å 19
       4.2.14 Intermolecular contacts less than 3.60 Å involving hydrogens 20

5. Crystal Images 26

5.1 Represents the ORTEP of the molecule (14) with thermal ellipsoids drawn at 50% probability 26

5.2 Represents the ORTEP of the molecule (14) with thermal ellipsoids drawn at 50% probability (only one molecule) 27

5.3 Packing diagram of the molecules when viewed down the b axis 28

5.4 Packing diagram of the molecules when viewed down the b axis (only one molecule) 29

5.5 Packing diagram of the molecules when viewed down the a axis 30

5.6 Packing diagram of the molecules when viewed down the a axis 31

6. References 32

Abbreviations

°C	Celsius
Å	Angstrom
°	Degree
mg	milligram
mmol	millimole
kV	Kilo volt
mA	milli ampere

Acknowledgements

First of all, I would like to thank Prof. Anamik Shah, *Vice-Chancellor, Gujarat Vidyapith, Ahmedabad and Department of Chemistry, Saurashtra University for giving me an opportunity to carry out work in his laboratory under his guidance. He shared with me a lot of his expertise and research insight. His wide knowledge in the field of Chemistry research and X-Ray logical way of thinking has been of great value for me to accomplish this chapter work. I am deeply grateful to the Institute for the trust and support that they gave me in order to study in Single Crystal X-Ray Instrumentation, SCX-mini model, Rigaku Company at Center of Excellence (CoE), National Facility for Drug Discovery Complex, Department of Chemistry, Saurashtra University, Rajkot, India. *This book is dedicated to my MOTHER "LINABEN RAMANLAL MEHARIYA" on her 50th Birthday.* I owe my special thanks to my beloved wife, Bhagwati Krunal Gauni Mehariya. My special gratitude is due to my parents, my sisters and their families for their loving support, inspiration to do my best in all matters of life. To them I dedicate this Chapter.

This book is also dedicated to my Grand Mother-Valiben Mehariya, Maniben Parmar, Mother in Law-Lataben Gauni and My Masi-Madhuben Mehariya.

Krunalkumar Mehariya

1. **Introduction**

   ❖ Oxazole is the aromatic compound of a wide range of heterocyclic aromatic organic scaffolds. Structures of Oxazole mainly contain a nitrogen atom and an oxygen atom separated by one carbon in a five membered ring [1].

   ❖ Oxazole is designated as 1, 3-oxazoles to indicate the position of heteroatoms in the ring. General structure of Oxazole (1) is given below.

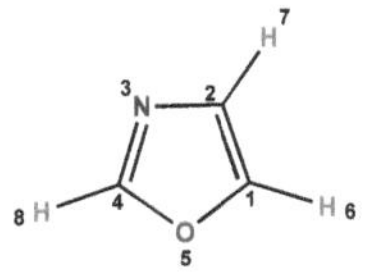

   Fig. 1 General Structure of Oxazole

   ❖ In the field of hetero cyclic chemistry, Oxazole and its derivatives are wide range and consist of medicinal chemistry, natural products, pharmaceutical chemistry and materials science.

   ❖ Oxazole heterocyclic ring of the proton acidities have been theoretically calculated and determined experimentally.

   ❖ The acidity of hydrogen atom decreases in the order C(4)<C(5) <C(2) carbon atom.

   ❖ Oxazoles have been found to be common skeleton in several naturally isolated compounds and have thus saved attention within the chemical and pharmaceutical public[3-4].

   ❖ The characteristic of oxazole skeleton checked by various types of the sophisticated analytical techniques.

   ❖ The IR spectrum of oxazole shows absorbance at 1143, 1080 (ring breathing), 1257 (C-H in plane deformation), 1536, 1499, 1327 (ring stretch), and 1045 cm^{-1}.

   ❖ The UV, the λ_{max} of Oxazole depends highly on the substitution pattern. In methanol, the parent ring system has an absorption maximum at λ_{max} ¼ 205 nm.

   ❖ Oxazole show characteristic resonances in both ^1H NMR and ^{13}C NMR spectra.

   ❖ The oxazole compound displays resonances (d) between 7.00 and 8.00 in the ^1H NMR spectrum, show that the aromatic region present in the skeleton, and the presence of substituent can change the chemical shift by up to 1 ppm[4-5].

2. **Oxazole-containing natural products**

   ❖ Several types of Oxazole-containing natural products were isolated from various regions and importance in pharmaceuticals as well as in synthetic organic fields.

   ❖ A number of oxazole skeletons are present in a great amount of natural products possessing a wide range of pharmacological activities. The past few years of the marine natural products with 2, 4-concatenated oxazoles have been isolated and synthesized.

   ❖ Indolyl-oxazoles derivatives of 5-(indol-3-yl)oxazoles compounds occurs in a variety of natural products and possess a wide range of biological importance.

   ❖ A **Pimprinine (2)**, Indolyl-oxazoles derivatives has a wide range of biological activities, from antibiotic and fungicidal effects to monoamide oxidase inhibition and anti-epilepsy [6].

   ❖ Indolyl-oxazoles derivatives, **Streptochlorin (3)** has a wide-ranging fungicidal activity and also an anti-proliferative agent. Streptochlorin has isolated from the fermentation broth of a marine actinomycete isolated from marine sediment [7].

   ❖ A series of the **Breitfussins A (4a)** and **B (4b)** are related heterocyclic compounds, containing Indolyl-oxazole aromatic ring in the molecules and originated from the marine organism *Thelia Breitfussi* [8].

- **Breitfussin A and Breitfussin B** is core of Indole and Oxazole ring with interesting targets in many ways.

3. Chemistry

3.1 Synthesis of 5-(3-methoxy-4-(prop-2-yn-1 yloxy) phenyl) oxazole

3.2 Preparation of 5-(3-methoxy-4-(prop-2-yn-1 yloxy) phenyl) oxazole (6)

The 3-methoxy-4-(prop-2-yn-1-yloxy) benzaldehyde (**5**) (191 mg, 1.0 mmol), TosMIC (196 mg, 1.0 mmol) and K_2CO_3 (150 mg, 1.2 mmol) were dissolved in methanol (2.0 mL). The solution was stirred at room temperature and reflux for 3-4 hours at 70-75 °C, until the starting materials were not detected by TLC. Then, the solid formed was filtered and washed with cold water (5 mL) used without further purification.

3.3 Analytical data

5-(3-methoxy-4-(prop-2-yn-1 yloxy)phenyl) oxazole(6): White solid; Yield 90%; mp 180-182 °C; MS: *m/z* 230.

4. X-ray Structure Report

4.1 Experimental

4.1.1 Data Collection

- ❖ A colorless prism crystal of $C_{13}H_{11}NO_3$ having approximate dimensions of 0.560 × 0.480 × 0.300 mm was mounted on a glass fiber.
- ❖ All measurements were made on a Rigaku SCX mini diffractometer using graphite monochromated Mo-Kα radiation.
- ❖ The crystal-to-detector distance was 52.00 mm. Cell constants and an orientation matrix for data collection corresponded to a primitive orthorhombic cell with dimensions: a = 17.421(4) Å, b = 18.581(4) Å, c = 6.882(2) Å, V = 2227.6(8) Å^3.
- ❖ For Z = 8 and F.W. = 229.23, the calculated density is 1.367 g/cm3. Based on the reflection conditions of:0kl: k+l = 2n and h0l: h = 2n.
- ❖ Packing considerations, a statistical analysis of intensity distribution, and the successful solution and refinement of the structure, the space group was determined to be:**Pna21 (#33)**.
- ❖ The data were collected at a temperature of 20 + 1 °C to a maximum 2θ value of 55.0°.
- ❖ A total of 540 oscillation images were collected. A sweep of data was done using ω oscillations from -120.0 to 60.0° in 1.0° steps.
- ❖ The exposure rate was 10.0 [sec./°]. The detector swing angle was -30.80°.
- ❖ A second sweep was performed using ω oscillations from -120.0 to 60.0° in 1.0° steps.
- ❖ The exposure rate was 10.0 [sec./°].
- ❖ The detector swing angle was -30.80°.
- ❖ Another sweep was performed using ω oscillations from -120.0 to 60.0° in 1.0° steps.
- ❖ The detector swing angle was -30.80°.
- ❖ The crystal-to-detector distance was 52.00 mm.
- ❖ Readout was performed in the 0.146 mm pixel mode.

4.1.2 Data Reduction

- ❖ The 20901 reflections that were collected, 5047 were unique (R_{int} = 0.0652), equivalent reflections were merged.
- ❖ Data were collected and processed using CrystalClear (Rigaku).
- ❖ The linear absorption coefficient, ω, for Mo-Kα radiation is 0.981 cm^{-1}.
- ❖ An empirical absorption correction was applied which resulted in transmission factors ranging from 0.635 to 0.971.
- ❖ The data were corrected for Lorentz and polarization effects.

4.1.3 Structure Solution and Refinement

- ❖ The structure was solved by direct methods [9] and expanded using Fourier techniques.
- ❖ The non-hydrogen atoms were refined anisotropically.
- ❖ Hydrogen atoms were refined using the riding model.
- ❖ The final cycle of full-matrix least-squares refinement [10]on F^2 was based on 5047 observed reflections and 319 variable parameters and converged (largest parameter shift was 1.19 times its esd) with unweighted and weighted agreement factors.
- ❖ The standard deviation of an observation of unit weight [11] was 0.96.
- ❖ Unit weights were used.
- ❖ The maximum and minimum peaks on the final difference Fourier map corresponded to 0.19 and -0.15 e/Å^3, respectively.
- ❖ The absolute structure was deduced based on Flack parameter, -3(4), using 2294 Friedel pairs [12].
- ❖ Neutral atom scattering factors were taken from Cromer and Waber [13].
- ❖ Anomalous dispersion effects were included in F_{calc}[14], the values for Δf' and Δf" were those of Creagh and McAuley [15].
- ❖ The values for the mass attenuation coefficients are those of Creagh and Hubbell [16].
- ❖ All calculations were performed using the CrystalStructure [17] crystallographic software package except for refinement, which was performed using SHELXL-97 [18-19].

4.2 Experimental Details

4.2.1 Crystal Data

Empirical Formula	$C_{13}H_{11}NO_3$
Formula Weight	229.23
Crystal Color, Habit	colorless, prism
Crystal Dimensions	0.560 ×0.480 × 0.300 mm
Crystal System	Orthorhombic
Lattice Type	Primitive
Lattice Parameters	a = 17.421(4) Å
	b = 18.581(4) Å
	c = 6.882(2) Å
	V = 2227.6(8) Å^3
Space Group	Pna2$_1$ (#33)
Z value	8
D$_{calc}$	1.367 g/cm^3
F$_{000}$	960.00
μ(MoKα)	0.981 cm^{-1}

4.2.2 Intensity Measurements

Diffractometer	SCX mini
Radiation	MoKα (λ = 0.71075 Å) graphite monochromated
Voltage, Current	50kV, 30mA
Temperature	20.0°C
Data Images	540 exposures
ω oscillation Range	-120.0 to 60.0°
Exposure Rate	10.0 sec./°
Detector Swing Angle	-30.80°
Detector Position	52.00 mm
Pixel Size	0.146 mm
$2\theta_{max}$	55.0°
No. of Reflections Measured	Total: 20901 Unique: 5047 (R_{int} = 0.0652) Friedel pairs: 2294
Corrections	Lorentz-polarization Absorption (trans. factors: 0.635 - 0.971)

4.2.3 Structure Solution and Refinement

Structure Solution	Direct Methods (SIR92)
Refinement	Full-matrix least-squares on F^2
Function Minimized	$\Sigma\omega\,(F_o^2 - F_c^2)^2$
Least Squares Weights	$\omega = 1/\,[\,\sigma^2(F_o^2) + (0.0455 \cdot P)^2 + 1.0896 \cdot P]$ where $P = (Max(F_o^2, 0) + 2F_c^2)/3$
2θ max cutoff	55.0°
Anomalous Dispersion	All non-hydrogen atoms
No. Observations (All reflections)	5047
No. Variables	319
Reflection/Parameter Ratio	15.82
Residuals: R_1 (I>2.00σ(I))	0.0538
Residuals: R (All reflections)	0.0947
Residuals: wR2 (All reflections)	0.1404
Goodness of Fit Indicator	0.961
Flack Parameter (Friedel pairs = 2294)	-3(4)
Max Shift/Error in Final Cycle	1.185
Maximum peak in Final Diff. Map	0.19 e /Å^3
Minimum peak in Final Diff. Map	-0.15 e /Å^3

4.2.4 Atomic coordinates and Biso/Beq

atom	x	y	z	Beq
O1	0.5294(3)	0.3280(2)	0.9875(6)	3.86(9)
O2	0.3506(2)	0.0393(2)	0.7626(7)	4.10(9)
O3	0.4932(2)	0.0146(2)	0.7552(7)	4.3(1)
O4	0.5299(2)	0.1717(2)	0.3498(6)	3.84(9)
O5	0.3503(2)	0.4590(2)	0.5757(7)	3.74(9)
O6	0.4945(2)	0.4841(2)	0.5829(7)	3.78(9)
N1	0.6558(3)	0.3352(3)	0.9665(9)	4.0(1)
N2	0.6561(3)	0.1659(3)	0.3589(9)	4.4(1)
C1	0.5948(3)	0.3670(3)	1.016(1)	3.8(1)
C2	0.6344(3)	0.2667(3)	0.922(1)	3.6(1)
C3	0.5568(3)	0.2622(3)	0.9309(8)	2.87(9)
C4	0.4999(3)	0.2058(3)	0.8911(9)	3.3(1)
C5	0.5252(3)	0.1386(3)	0.8403(8)	3.2(1)
C6	0.4744(3)	0.0837(3)	0.8027(9)	3.2(1)
C7	0.3972(3)	0.0978(3)	0.8069(9)	3.2(1)
C8	0.3713(3)	0.1660(3)	0.852(1)	3.7(2)
C9	0.4215(3)	0.2197(4)	0.896(1)	3.9(1)
C10	0.5720(3)	-0.0049(3)	0.769(2)	4.3(2)
C11	0.2744(3)	0.0534(4)	0.709(1)	3.8(2)
C12	0.2674(3)	0.0902(3)	0.526(1)	3.9(2)
C13	0.2609(4)	0.1182(4)	0.373(1)	5.2(2)
C14	0.5924(4)	0.1319(3)	0.334(1)	4.5(2)
C15	0.6289(4)	0.2332(3)	0.424(1)	4.0(1)
C16	0.5538(4)	0.2377(3)	0.4128(8)	3.4(1)
C17	0.4996(3)	0.2929(3)	0.4502(8)	2.86(9)
C18	0.5256(3)	0.3628(3)	0.4928(9)	3.1(1)
C19	0.4749(3)	0.4157(3)	0.5383(9)	2.93(9)
C20	0.3951(3)	0.4014(3)	0.537(1)	3.15(9)
C21	0.3699(3)	0.3355(3)	0.486(1)	3.8(2)
C22	0.4229(3)	0.2806(3)	0.442(1)	3.6(1)
C23	0.2719(3)	0.4443(4)	0.629(1)	4.2(2)
C24	0.2671(4)	0.4092(5)	0.819(2)	4.9(2)
C25	0.2618(5)	0.3825(5)	0.966(2)	7.4(3)
C32	0.5720(3)	0.5040(4)	0.573(2)	4.8(2)

Beq = 8/3 π^2(U11(aa*)^2 + U22(bb*)^2 + U33(cc*)^2 + 2U12(aa*bb*)cos γ + 2U13(aa*cc*)cosβ + 2U23(bb*cc*)cos α)

4.2.5 Atomic coordinates and Biso involving hydrogen atoms

atom	x	y	z	B_{iso}
H1	0.5946	0.4133	1.0675	4.52
H2	0.672(2)	0.230(2)	0.902(5)	3.9(8)
H5	0.5776	0.1300	0.8311	3.81
H8	0.3189	0.1754	0.8529	4.45
H9	0.4035	0.2652	0.9296	4.63
H10A	0.5891	0.0009	0.9006	5.12
H10B	0.6019	0.0256	0.6852	5.12
H10C	0.5783	-0.0541	0.7302	5.12
H11A	0.2506	0.0825	0.8098	4.54
H11B	0.2466	0.0083	0.7021	4.54
H13	0.2557	0.1401	0.2519	6.29
H14	0.5905	0.0829	0.3067	5.44
H15	0.6603	0.2699	0.4699	4.81
H18	0.5779	0.3728	0.4897	3.66
H21	0.3175	0.3261	0.4805	4.59
H22	0.4051	0.2353	0.4069	4.31
H23A	0.254(2)	0.498(2)	0.649(5)	3.4(7)
H23B	0.241(2)	0.417(2)	0.525(6)	4.5(8)
H25	0.2575	0.3605	1.0873	8.82
H32A	0.5904	0.4978	0.4426	5.75
H32B	0.6015	0.4743	0.6595	5.75

4.2.6 Anisotropic displacement parameters

atom	U11	U22	U33	U12	U13	U23
O1	0.055(3)	0.037(2)	0.055(3)	-0.002(2)	-0.009(2)	-0.003(2)
O2	0.034(2)	0.047(3)	0.075(4)	-0.002(2)	-0.022(2)	0.000(3)
O3	0.035(2)	0.047(3)	0.082(4)	0.009(2)	-0.002(3)	-0.008(3)
O4	0.050(3)	0.042(2)	0.054(3)	0.001(2)	-0.012(2)	-0.004(2)
O5	0.042(2)	0.037(2)	0.063(3)	-0.002(2)	-0.001(2)	0.004(2)
O6	0.043(2)	0.034(2)	0.066(3)	0.000(2)	0.009(2)	-0.011(2)
N1	0.045(3)	0.051(3)	0.055(3)	-0.004(2)	-0.007(3)	0.001(3)
N2	0.065(3)	0.044(3)	0.058(4)	0.011(3)	-0.011(3)	-0.006(3)
C1	0.053(4)	0.048(3)	0.043(4)	-0.013(3)	0.004(3)	-0.005(3)
C2	0.036(3)	0.054(3)	0.048(3)	-0.005(3)	-0.008(3)	-0.002(3)
C3	0.039(3)	0.036(3)	0.034(3)	-0.004(2)	-0.010(3)	-0.001(3)
C4	0.044(3)	0.045(3)	0.039(3)	-0.002(2)	-0.011(3)	0.006(3)
C5	0.033(3)	0.053(3)	0.034(3)	-0.001(3)	-0.001(3)	-0.003(3)
C6	0.046(3)	0.030(3)	0.044(3)	0.003(3)	-0.008(3)	0.005(3)
C7	0.041(3)	0.039(3)	0.041(3)	-0.003(3)	-0.017(3)	0.009(3)
C8	0.039(3)	0.050(4)	0.052(4)	0.002(3)	-0.008(3)	0.002(3)
C9	0.038(3)	0.064(4)	0.045(4)	0.003(3)	-0.004(3)	0.001(3)
C10	0.048(4)	0.041(3)	0.073(5)	0.018(3)	0.002(4)	-0.000(4)

C11	0.035(4)	0.046(4)	0.062(4)	-0.003(3)	0.001(3)	-0.009(3)
C12	0.045(3)	0.041(3)	0.064(4)	0.015(3)	-0.022(3)	-0.011(3)
C13	0.079(5)	0.067(5)	0.053(4)	0.025(4)	-0.034(4)	-0.007(4)
C14	0.073(5)	0.032(3)	0.067(5)	0.010(3)	0.010(4)	0.002(3)
C15	0.058(4)	0.034(3)	0.061(4)	-0.003(3)	-0.008(3)	-0.008(3)
C16	0.061(4)	0.035(3)	0.032(3)	-0.007(3)	-0.006(3)	0.003(3)
C17	0.044(3)	0.038(3)	0.027(3)	-0.000(3)	-0.002(3)	-0.000(3)
C18	0.036(3)	0.036(3)	0.043(4)	-0.001(3)	0.005(3)	0.004(3)
C19	0.037(3)	0.044(3)	0.030(3)	0.001(3)	-0.002(3)	-0.002(3)
C20	0.037(3)	0.039(3)	0.043(3)	0.001(2)	-0.005(3)	-0.001(3)
C21	0.035(3)	0.048(3)	0.062(5)	-0.009(3)	0.001(3)	-0.000(3)
C22	0.058(3)	0.020(2)	0.059(4)	-0.006(2)	0.005(3)	-0.005(3)
C23	0.031(3)	0.066(5)	0.061(4)	-0.003(3)	0.009(3)	-0.004(4)
C24	0.045(4)	0.075(5)	0.068(5)	-0.003(3)	-0.001(4)	-0.012(5)
C25	0.102(7)	0.086(6)	0.091(7)	-0.022(5)	-0.011(5)	0.005(6)
C32	0.035(3)	0.069(4)	0.079(5)	-0.002(3)	0.005(3)	-0.022(4)

The general temperature factor expression: $\exp(-2\pi 2(a^{*2}U_{11}h^2 + b^{*2}U_{22}k^2 + c^{*2}U_{33}l^2 + 2a^*b^*U_{12}hk + 2a^*c^*U_{13}hl + 2b^*c^*U_{23}kl))$

4.2.7 Bond lengths (Å)

atom	atom	distance	atom	atom	distance
O1	C1	1.366(7)	O1	C3	1.369(6)
O2	C7	1.391(7)	O2	C11	1.401(7)
O3	C6	1.364(6)	O3	C10	1.423(6)
O4	C14	1.321(7)	O4	C16	1.365(7)
O5	C20	1.350(7)	O5	C23	1.441(7)
O6	C19	1.351(6)	O6	C32	1.403(6)
N1	C1	1.261(8)	N1	C2	1.360(8)
N2	C14	1.289(8)	N2	C15	1.411(8)
C2	C3	1.357(8)	C3	C4	1.467(7)
C4	C5	1.369(8)	C4	C9	1.392(7)
C5	C6	1.376(8)	C6	C7	1.371(8)
C7	C8	1.380(8)	C8	C9	1.360(8)
C11	C12	1.439(10)	C12	C13	1.183(10)
C15	C16	1.312(9)	C16	C17	1.419(8)
C17	C18	1.406(7)	C17	C22	1.356(7)
C18	C19	1.359(7)	C19	C20	1.416(7)
C20	C21	1.346(8)	C21	C22	1.408(7)
C23	C24	1.461(12)	C24	C25	1.135(14)

4.2.8 Bond lengths involving hydrogens (Å)

atom	atom	distance	atom	atom	distance
C1	H1	0.930	C2	H2	0.95(4)
C5	H5	0.930	C8	H8	0.930
C9	H9	0.930	C10	H10A	0.960
C10	H10B	0.960	C10	H10C	0.960
C11	H11A	0.970	C11	H11B	0.970
C13	H13	0.930	C14	H14	0.930
C15	H15	0.930	C18	H18	0.930
C21	H21	0.930	C22	H22	0.930
C23	H23A	1.05(4)	C23	H23B	1.02(4)
C25	H25	0.930	C32	H32A	0.960
C32	H32B	0.960	C32	H32C	0.960

4.2.9 Bond angles (°)

atom	atom	atom	angle	atom	atom	atom	angle	Atom
C1	O1	C3	102.9(4)	C7	O2	C11	117.5(4)	C1
C6	O3	C10	117.0(4)	C14	O4	C16	106.1(5)	C6
C20	O5	C23	116.6(5)	C19	O6	C32	118.7(4)	C20
C1	N1	C2	105.6(5)	C14	N2	C15	100.8(5)	C1
O1	C1	N1	114.5(5)	N1	C2	C3	108.7(5)	O1
O1	C3	C2	107.7(5)	O1	C3	C4	117.1(5)	O1
C2	C3	C4	135.1(5)	C3	C4	C5	118.8(5)	C2
C3	C4	C9	121.7(5)	C5	C4	C9	119.4(5)	C3
C4	C5	C6	121.2(5)	O3	C6	C5	126.1(5)	C4
O3	C6	C7	114.9(5)	C5	C6	C7	119.0(5)	O3
O2	C7	C6	114.7(5)	O2	C7	C8	125.3(5)	O2
C6	C7	C8	120.1(5)	C7	C8	C9	120.9(5)	C6
C4	C9	C8	119.3(6)	O2	C11	C12	113.5(5)	C4
C11	C12	C13	177.6(7)	O4	C14	N2	115.1(5)	C11
N2	C15	C16	111.8(5)	O4	C16	C15	105.5(5)	N2
O4	C16	C17	120.2(5)	C15	C16	C17	134.3(5)	O4
C16	C17	C18	119.4(5)	C16	C17	C22	121.8(5)	C16
C18	C17	C22	118.8(5)	C17	C18	C19	120.4(5)	C18
O6	C19	C18	124.7(5)	O6	C19	C20	115.3(5)	O6
C18	C19	C20	120.1(5)	O5	C20	C19	114.6(5)	C18
O4	C16	C17	120.2(5)	C15	C16	C17	134.3(5)	O4
O5	C20	C21	125.8(5)	C19	C20	C21	119.5(5)	O5
C20	C21	C22	120.1(5)	C17	C22	C21	121.0(5)	C20
O5	C23	C24	111.5(5)	C23	C24	C25	178.6(8)	O5

4.2.10 Bond angles involving hydrogens (°)

atom	atom	atom	angle	atom	atom	atom	angle	Atom
O1	C1	H1	122.7	N1	C1	H1	122.8	O1
N1	C2	H2	121(2)	C3	C2	H2	130(3)	N1
C4	C5	H5	119.4	C6	C5	H5	119.4	C4
C7	C8	H8	119.6	C9	C8	H8	119.5	C7
C4	C9	H9	120.3	C8	C9	H9	120.3	C4
O3	C10	H10A	109.5	O3	C10	H10B	109.5	O3
H10A	C10	H10C	109.5	H10B	C10	H10C	109.5	H10A
O2	C11	H11A	108.9	O2	C11	H11B	108.9	O2
C12	C11	H11A	108.9	C12	C11	H11B	108.9	C12
H11A	C11	H11B	107.7	C12	C13	H13	180.0	H11A
O4	C14	H14	122.5	N2	C14	H14	122.4	O4
N2	C15	H15	124.1	C16	C15	H15	124.1	N2
C17	C18	H18	119.8	C19	C18	H18	119.8	C17
C20	C21	H21	120.0	C22	C21	H21	120.0	C20
C17	C22	H22	119.5	C21	C22	H22	119.5	C17
O5	C23	H23A	97.9(17)	O5	C23	H23B	114(2)	O5
C24	C23	H23A	106.8(19)	C24	C23	H23B	112(2)	C24
H23A	C23	H23B	114(3)	C24	C25	H25	180.0	H23A
O6	C32	H32A	109.5	O6	C32	H32B	109.5	O6
O6	C32	H32C	109.5	H32A	C32	H32B	109.5	O6
H32A	C32	H32C	109.5	H32B	C32	H32C	109.5	H32A

4.2.11 Torsion Angles (°)
(Those having bond angles > 160 or < 20 degrees are excluded.)

atom^1	atom^2	atom^3	atom^4	Angle
C1	O1	C3	C2	-2.1(6)
C1	O1	C3	C4	179.0(5)
C3	O1	C1	N1	6.3(7)
C7	O2	C11	C12	67.2(7)
C11	O2	C7	C6	-162.0(5)
C11	O2	C7	C8	17.9(9)
C10	O3	C6	C5	8.2(9)
C10	O3	C6	C7	-173.5(5)
C14	O4	C16	C15	1.9(6)
C14	O4	C16	C17	-178.1(5)
C16	O4	C14	N2	-7.2(8)
C20	O5	C23	C24	68.4(7)
C23	O5	C20	C19	-163.2(5)
C23	O5	C20	C21	21.1(9)
C32	O6	C19	C18	3.3(9)

C32	O6	C19	C20	-175.2(5)
C1	N1	C2	C3	5.8(7)
C2	N1	C1	O1	-7.6(7)
C14	N2	C15	C16	-7.2(7)
C15	N2	C14	O4	8.7(8)
N1	C2	C3	O1	-2.1(7)
N1	C2	C3	C4	176.5(6)
O1	C3	C4	C5	-177.9(5)
O1	C3	C4	C9	4.6(8)
C2	C3	C4	C5	3.6(10)
C2	C3	C4	C9	-173.9(6)
C3	C4	C5	C6	179.6(5)
C3	C4	C9	C8	178.0(5)
C5	C4	C9	C8	0.5(9)
C9	C4	C5	C6	-2.9(9)
C4	C5	C6	O3	-178.5(5)
C4	C5	C6	C7	3.3(9)
O3	C6	C7	O2	0.2(8)
O3	C6	C7	C8	-179.7(5)
C5	C6	C7	O2	178.6(5)
C5	C6	C7	C8	-1.3(9)
O2	C7	C8	C9	179.1(5)
C6	C7	C8	C9	-1.0(9)
C7	C8	C9	C4	1.4(10)
N2	C15	C16	O4	3.4(7)
N2	C15	C16	C17	-176.6(6)
O4	C16	C17	C18	-172.6(5)
O4	C16	C17	C22	6.0(8)
C15	C16	C17	C18	7.4(10)
C15	C16	C17	C22	-174.0(6)
C16	C17	C18	C19	-176.8(5)
C16	C17	C22	C21	177.8(5)
C18	C17	C22	C21	-3.6(9)
C22	C17	C18	C19	4.6(8)
C17	C18	C19	O6	179.4(5)
C17	C18	C19	C20	-2.2(8)
O6	C19	C20	O5	1.4(8)
O6	C19	C20	C21	177.4(5)
C18	C19	C20	O5	-177.1(5)
C18	C19	C20	C21	-1.2(9)
O5	C20	C21	C22	177.6(5)
C19	C20	C21	C22	2.2(9)
C20	C21	C22	C17	0.3(10)

4.2.12 Intramolecular contacts less than 3.60 Å involving hydrogens

atom	atom	distance	atom	atom	Distance
O1	H2	3.13(4)	O1	H9	2.516
O2	H8	2.662	O3	H5	2.652
O4	H15	3.030	O4	H22	2.505
O5	H21	2.617	O6	H18	2.607
C1	H2	2.98(4)	C2	H1	2.983
C2	H5	2.797	C3	H1	3.033
C3	H5	2.576	C3	H9	2.671
C4	H2	3.03(4)	C4	H8	3.216
C5	H2	3.10(4)	C5	H9	3.225
C5	H10A	2.821	C5	H10B	2.709
C6	H8	3.219	C6	H10A	2.610
C6	H10B	2.599	C6	H10C	3.175
C7	H5	3.205	C7	H9	3.224
C7	H11A	2.569	C7	H11B	3.188
C8	H11A	2.629	C9	H5	3.222
C10	H5	2.543	C11	H8	2.589
C11	H13	3.551	C12	H8	2.891
C13	H11A	3.086	C13	H11B	3.061
C14	H15	2.976	C15	H14	2.983
C15	H18	2.779	C16	H14	3.035
C16	H18	2.601	C16	H22	2.592
C17	H15	2.836	C17	H21	3.238
C18	H15	2.917	C18	H22	3.220
C18	H32A	2.773	C18	H32B	2.713
C19	H21	3.232	C19	H32A	2.609
C19	H32B	2.598	C19	H32C	3.161
C20	H18	3.246	C20	H22	3.218
C20	H23A	3.14(3)	C20	H23B	2.69(4)
C21	H23B	2.72(4)	C22	H18	3.214
C23	H21	2.549	C23	H25	3.525
C24	H21	2.928	C25	H23A	3.06(4)
C25	H23B	3.13(4)	C32	H18	2.507
H2	H5	2.529	H5	H10A	2.453
H5	H10B	2.225	H5	H10C	3.491
H8	H9	2.289	H8	H11A	2.116
H8	H11B	3.506	H15	H18	2.395
H18	H32A	2.356	H18	H32B	2.257
H18	H32C	3.459	H21	H22	2.331
H21	H23A	3.571	H21	H23B	2.171

4.2.13 Intermolecular contacts less than 3.60 Å

atom	atom	distance	atom	atom	distance
O1	O6^1	3.578(5)	O1	C16^2	3.401(7)
O1	C17^2	3.292(7)	O1	C18	3.466(7)
O1	C18^2	3.537(7)	O2	C14^3	3.367(7)
O3	O34	3.491(7)	O3	O33	3.491(7)
O3	O43	3.545(5)	O3	C10^4	3.538(9)
O3	C14^3	3.150(7)	O4	O3^4	3.545(5)
O4	C3^5	3.369(7)	O4	C4^5	3.261(7)
O4	C55	3.561(7)	O4	C5	3.432(7)
O5	C16	3.397(7)	O6	O1^6	3.578(5)
O6	O66	3.497(7)	O6	O61	3.497(7)
O6	C16	3.208(7)	O6	C32^1	3.574(9)
N1	C11^7	3.419(8)	N1	C13^8	3.452(9)
N2	C2^5	3.559(9)	N2	C23^7	3.423(9)
N2	C25^9	3.391(12)	C1	O51	3.397(7)
C1	O61	3.208(7)	C1	C18^2	3.497(9)
C2	N22	3.559(9)	C2	C15	3.486(10)
C2	C15^2	3.510(10)	C2	C25^7	3.563(10)
C3	O42	3.369(7)	C3	C16	3.595(8)
C3	C16^2	3.347(8)	C3	C17	3.502(8)
C3	C18	3.589(8)	C4	O42	3.261(7)
C4	C16	3.474(8)	C4	C17	3.439(8)
C5	O4	3.432(7)	C5	O42	3.561(7)
C5	C14^2	3.595(9)	C5	C16	3.506(8)
C9	C22	3.325(9)	C10	O33	3.538(9)
C11	N1^{10}	3.419(8)	C13	N1^{11}	3.452(9)
C14	O2^4	3.367(7)	C14	O34	3.150(7)
C14	C5^5	3.595(9)	C15	C25	3.510(10)
C15	C2	3.486(10)	C16	O1^5	3.401(7)
C16	C3^5	3.347(8)	C16	C3	3.595(8)
C16	C4	3.474(8)	C16	C5	3.506(8)
C17	O1^5	3.292(7)	C17	C3	3.502(8)
C17	C4	3.439(8)	C18	O1^5	3.537(7)
C18	O1	3.466(7)	C18	C1^5	3.497(9)
C18	C3	3.589(8)	C22	C9	3.325(9)
C23	N2^{10}	3.423(9)	C25	N2^{12}	3.391(12)
C25	C2^{10}	3.563(10)	C32	O6^6	3.574(9)

Symmetry Operators:

(1) -X+1,-Y+1,Z+1/2

(2) X,Y,Z+1

(3) -X+1,-Y,Z+1/2

(4) -X+1,-Y,Z+1/2-1

(5) X,Y,Z-1

(6) -X+1,-Y+1,Z+1/2-1

(7) X+1/2,-Y+1/2,Z

(8) X+1/2,-Y+1/2,Z+1

(9) X+1/2,-Y+1/2,Z-1

(10) X+1/2-1,-Y+1/2,Z

(11) X+1/2-1,-Y+1/2,Z-1

(12) X+1/2-1,-Y+1/2,Z+1

4.2.14 Intermolecular contacts less than 3.60 Å involving hydrogens

atom	atom	distance	atom	atom	distance
O1	H32C^1	3.004	O2	H10A^2	2.805
O2	H10B^3	3.254	O2	H10C^3	3.459
O2	H14^3	2.509	O2	H23A^4	3.31(4)
O2	H23B^4	3.31(4)	O3	H10A^2	2.844
O3	H10B^3	3.472	O3	H10C^3	3.575
O3	H14^3	2.352	O4	H5	3.502
O4	H10C^2	3.000	O5	H1^5	2.561
O5	H11A^6	3.422	O5	H11B^6	3.210
O5	H32A^1	2.844	O5	H32B^5	3.231
O5	H32C^5	3.448	O6	H1^5	2.460
O6	H32A^1	2.902	O6	H32B^5	3.447
O6	H32C^5	3.553	N1	H8^7	2.954
N1	H11A^7	2.497	N1	H13^8	2.665
N1	H32B	3.470	N2	H2^9	3.37(4)
N2	H5	3.588	N2	H10B	3.569
N2	H21^7	2.938	N2	H23B^7	2.43(4)
N2	H25^{10}	2.618	C1	H11A^7	3.203
C1	H13^8	3.242	C1	H18^{11}	3.275

C1	H32B	3.163	C1	H32C^1	3.406
C2	H8^7	3.423	C2	H11A^7	3.543
C2	H13^8	3.550	C2	H15	3.147
C2	H25^7	3.387	C5	H10C^3	3.593
C5	H14^{11}	3.560	C6	H10A^2	3.368
C6	H10B^3	3.579	C6	H10C^3	3.131
C6	H14^3	3.294	C7	H10A^2	3.353
C7	H10B^3	3.469	C7	H10C^3	3.054
C7	H14^3	3.364	C8	H10C^3	3.443
C8	H13^{11}	3.442	C8	H22	3.376
C9	H22	3.392	C9	H22^{11}	3.537
C10	H14	3.588	C10	H14^3	3.192
C10	H23A^7	3.28(3)	C11	H10A^2	3.344
C11	H14^3	3.521	C11	H23A^4	3.23(4)
C11	H23B^4	3.35(4)	C11	H32B^{12}	3.076
C12	H10A^2	3.141	C12	H10C^2	3.438
C12	H15^{12}	3.222	C12	H18^{12}	3.382
C12	H23A^{13}	3.14(4)	C12	H32A^{12}	3.539
C12	H32B^{12}	3.261	C13	H10A^2	3.430
C7	H10B^3	3.469	C7	H10C^3	3.054
C7	H14^3	3.364	C8	H10C^3	3.443
C8	H13^{11}	3.442	C8	H22	3.376
C9	H22	3.392	C9	H22^{11}	3.537
C10	H14	3.588	C10	H14^3	3.192
C10	H23A^7	3.28(3)	C11	H10A^2	3.344
C11	H14^3	3.521	C11	H23A^4	3.23(4)
C11	H23B^4	3.35(4)	C11	H32B^{12}	3.076
C12	H10A^2	3.141	C12	H10C^2	3.438
C12	H15^{12}	3.222	C12	H18^{12}	3.382
C12	H23A^{13}	3.14(4)	C12	H32A^{12}	3.539
C13	H10C^2	3.198	C13	H15^{12}	2.800
C13	H18^{12}	3.292	C13	H22	3.332
C13	H23A^{13}	2.73(4)	C14	H5^9	3.470
C14	H5	3.431	C14	H10B	3.126
C14	H10C^2	3.382	C14	H23B^7	3.05(4)
C14	H25^{10}	3.343	C15	H2	3.37(4)
C15	H5	3.509	C15	H13^7	3.439
C15	H21^7	3.488	C15	H23B^7	3.48(4)
C16	H5	3.530	C18	H1^9	3.299
C18	H32C^5	3.544	C19	H1^5	3.406
C19	H32A^1	3.407	C19	H32B^5	3.569
C19	H32C^5	3.134	C20	H1^5	3.455
C20	H32A^1	3.373	C20	H32B^5	3.474

C20	H32C^5	3.087	C21	H9	3.370
C21	H25^9	3.404	C21	H32C^5	3.432
C22	H9^9	3.555	C22	H9	3.384
C23	H1^5	3.550	C23	H10B^{12}	3.039
C23	H11A^6	3.402	C23	H11B^6	3.186
C23	H32A^1	3.402	C24	H2^{12}	3.13(4)
C24	H5^{12}	3.381	C24	H10A^{12}	3.566
C24	H10B^{12}	3.255	C24	H11B^{14}	3.226
C24	H32A^1	3.143	C24	H32C^1	3.443
C25	H2^{12}	2.65(4)	C25	H5^{12}	3.349
C25	H9	3.303	C25	H11B^{14}	2.850
C25	H32A^1	3.407	C25	H32C^1	3.201
C32	H1^5	3.284	C32	H11B^7	3.177
H1	O5^1	2.561	H1	O6^1	2.460
H1	C18^{11}	3.299	H1	C19^1	3.406
H1	C20^1	3.455	H1	C23^1	3.550
H1	C32^1	3.284	H1	H11A^7	3.247
H1	H13^8	3.237	H1	H18^{11}	3.015
H1	H23A^1	3.165	H1	H32A^{11}	3.023
H1	H32B	3.031	H1	H32C^1	3.072
H2	N2^{11}	3.37(4)	H2	C15	3.37(4)
H2	C24^7	3.13(4)	H2	C25^7	2.65(4)
H2	H8^7	3.124	H2	H15	3.070
H2	H25^7	2.588	H5	O4	3.502
H5	N2	3.588	H5	C14	3.431
H5	C14^{11}	3.470	H5	C15	3.509
H5	C16	3.530	H5	C24^7	3.381
H5	C25^7	3.349	H5	H14^{11}	3.395
H5	H25^7	3.600	H8	N1^{12}	2.954
H8	C2^{12}	3.423	H8	H2^{12}	3.124
H8	H13^{11}	3.030	H8	H22	3.593
H9	C21	3.370	H9	C22	3.384
H9	C22^{11}	3.555	H9	C25	3.303
H5	N2	3.588	H5	C14	3.431
H5	C14^{11}	3.470	H5	C15	3.509
H5	C16	3.530	H5	C24^7	3.381
H5	C25^7	3.349	H5	H14^{11}	3.395
H5	H25^7	3.600	H8	N1^{12}	2.954
H8	C2^{12}	3.423	H8	H2^{12}	3.124
H8	H13^{11}	3.030	H8	H22	3.593
H9	C21	3.370	H9	C22	3.384
H9	H22^{11}	3.332	H9	H25	3.285
H10A	O2^3	2.805	H10A	O3^3	2.844

H10A	C6^3	3.368	H10A	C7^3	3.353
H10A	C11^3	3.344	H10A	C12^3	3.141
H10A	C13^3	3.430	H10A	C24^7	3.566
H10A	H11B^3	3.538	H10A	H14^{11}	3.183
H10A	H14^3	3.555	H10A	H23A^7	3.352
H10B	O2^2	3.254	H10B	O3^2	3.472
H10B	N2	3.569	H10B	C6^2	3.579
H10B	C7^2	3.469	H10B	C14	3.126
H10B	C23^7	3.039	H10B	C24^7	3.255
H10B	H14	2.820	H10B	H23A^7	2.694
H10B	H23B^7	2.875	H10C	O2^2	3.459
H10C	O3^2	3.575	H10C	O4^3	3.000
H10C	C5^2	3.593	H10C	C6^2	3.131
H10C	C7^2	3.054	H10C	C8^2	3.443
H10C	C12^3	3.438	H10C	C13^3	3.198
H10C	C14^3	3.382	H10C	H13^3	3.307
H10C	H14^3	3.035	H10C	H22^3	3.590
H10C	H23A^7	3.279	H11A	O5^4	3.422
H11A	N1^{12}	2.497	H11A	C1^{12}	3.203
H11A	C2^{12}	3.543	H11A	C23^4	3.402
H11A	H1^{12}	3.247	H11A	H13^{11}	3.227
H11A	H23A^4	2.814	H11A	H23B^4	3.412
H11A	H32B^{12}	2.989	H11B	O5^4	3.210
H11B	C23^4	3.186	H11B	C24^{13}	3.226
H11B	C25^{13}	2.850	H11B	C32^{12}	3.177
H11B	H10A^2	3.538	H11B	H14^3	3.382
H10C	H23A^7	3.279	H11A	O5^4	3.422
H11A	N1^{12}	2.497	H11A	C1^{12}	3.203
H11A	C2^{12}	3.543	H11A	C23^4	3.402
H11A	H1^{12}	3.247	H11A	H13^{11}	3.227
H11A	H23A^4	2.814	H11A	H23B^4	3.412
H11A	H32B^{12}	2.989	H11B	O5^4	3.210
H11B	C23^4	3.186	H11B	C24^{13}	3.226
H11B	C25^{13}	2.850	H11B	C32^{12}	3.177
H11B	H10A^2	3.538	H11B	H14^3	3.382
H11B	H23A^4	3.080	H11B	H23B^4	2.800
H11B	H25^{13}	2.858	H11B	H32A^{12}	3.258
H11B	H32B^{12}	2.566	H11B	H32C^{12}	3.227
H13	N1^{15}	2.665	H13	C1^{15}	3.242
H13	C2^{15}	3.550	H13	C8^9	3.442
H13	C15^{12}	3.439	H13	H1^{15}	3.237
H13	H8^9	3.030	H13	H10C^2	3.307
H13	H11A^9	3.227	H13	H15^{12}	2.794

H13	H18^{12}	3.512	H13	H22	3.322
H13	H23A^{13}	2.742	H14	O2^{2}	2.509
H14	O3^{2}	2.352	H14	C5^{9}	3.560
H14	C6^{2}	3.294	H14	C7^{2}	3.364
H14	C10	3.588	H14	C10^{2}	3.192
H14	C11^{2}	3.521	H14	H5^{9}	3.395
H14	H10A^{9}	3.183	H14	H10A^{2}	3.555
H14	H10B	2.820	H14	H10C^{2}	3.035
H14	H11B^{2}	3.382	H14	H23B^{7}	3.028
H14	H25^{10}	3.442	H15	C2	3.147
H15	C12^{7}	3.222	H15	C13^{7}	2.800
H15	H2	3.070	H15	H13^{7}	2.794
H15	H21^{7}	3.270	H18	C1^{9}	3.275
H18	C12^{7}	3.382	H18	C13^{7}	3.292
H18	H1^{9}	3.015	H18	H13^{7}	3.512
H21	N2^{12}	2.938	H21	C15^{12}	3.488
H21	H15^{12}	3.270	H21	H25^{9}	2.971
H22	C8	3.376	H22	C9^{9}	3.537
H22	C9	3.392	H22	C13	3.332
H22	H8	3.593	H22	H9^{9}	3.332
H22	H10C^{2}	3.590	H22	H13	3.322
H23A	O2^{6}	3.31(4)	H23A	C10^{12}	3.28(3)
H23A	C11^{6}	3.23(4)	H23A	C12^{14}	3.14(4)
H23A	C13^{14}	2.73(4)	H23A	H1^{5}	3.165
H23B	N2^{12}	2.43(4)	H23B	C11^{6}	3.35(4)
H23B	C14^{12}	3.05(4)	H23B	C15^{12}	3.48(4)
H23B	H10B^{12}	2.875	H23B	H11A^{6}	3.412
H23B	H11B^{6}	2.800	H23B	H14^{12}	3.028
H23B	H25^{9}	3.202	H25	N2^{16}	2.618
H25	C2^{12}	3.387	H25	C14^{16}	3.343
H25	C21^{11}	3.404	H25	H2^{12}	2.588
H25	H5^{12}	3.600	H25	H9	3.285
H25	H11B^{14}	2.858	H25	H14^{16}	3.442
H25	H21^{11}	2.971	H25	H23B^{14}	3.202
H25	H32C^{1}	3.294	H32A	O5^{5}	2.844
H32A	O6^{5}	2.902	H32A	C12^{7}	3.539
H32A	C19^{5}	3.407	H32A	C20^{5}	3.373
H32A	C23^{5}	3.402	H32A	C24^{5}	3.143
H32A	C25^{5}	3.407	H32A	H1^{9}	3.023
H32A	H11B^{7}	3.258	H32A	H23A^{5}	3.386
H32B	O5^{1}	3.231	H32B	O6^{1}	3.447
H32B	N1	3.470	H32B	C1	3.163
H32B	C11^{7}	3.076	H32B	C12^{7}	3.261

H32B	C19[1]	3.569		H32B	C20[1]	3.474
H32B	H1	3.031		H32B	H11A[7]	2.989
H32B	H11B[7]	2.566		H32C	O1[5]	3.004
H32C	O5[1]	3.448		H32C	O6[1]	3.553
H32C	C1[5]	3.406		H32C	C18[1]	3.544
H32C	C19[1]	3.134		H32C	C20[1]	3.087
H32C	C21[1]	3.432		H32C	C24[5]	3.443
H32C	C25[5]	3.201		H32C	H1[5]	3.072
H32C	H11B[7]	3.227		H32C	H25[5]	3.294

Symmetry Operators:

(1) -X+1,-Y+1,Z+1/2

(2) -X+1,-Y,Z+1/2-1

(3) -X+1,-Y,Z+1/2

(4) -X+1/2,Y+1/2-1,Z+1/2

(5) -X+1,-Y+1,Z+1/2-1

(6) -X+1/2,Y+1/2,Z+1/2-1

(7) X+1/2,-Y+1/2,Z

(8) X+1/2,-Y+1/2,Z+1

(9) X,Y,Z-1

(10) X+1/2,-Y+1/2,Z-1

(11) X,Y,Z+1

(12) X+1/2-1,-Y+1/2,Z

(13) -X+1/2,Y+1/2-1,Z+1/2-1

(14) -X+1/2,Y+1/2,Z+1/2

(15) X+1/2-1,-Y+1/2,Z-1

(16) X+1/2-1,-Y+1/2,Z+1

5. Crystal Images

5.1 Represents the ORTEP of the molecule (14) with thermal ellipsoids drawn at 50% probability

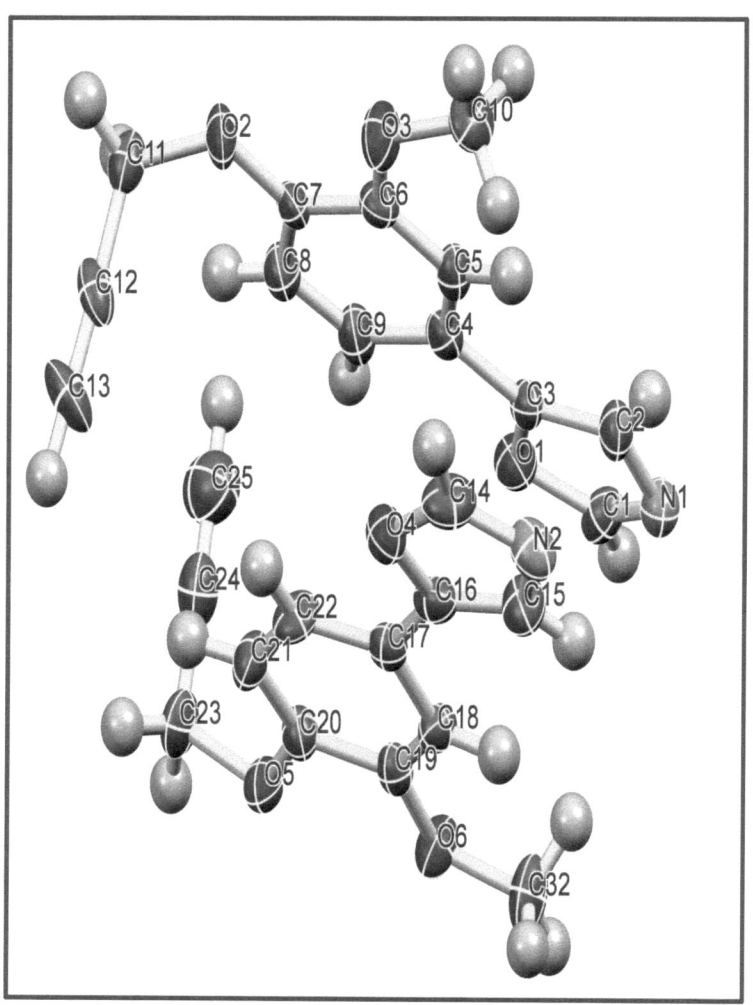

5.2 Represents the ORTEP of the molecule (14) with thermal ellipsoids drawn at 50% probability (only one molecule)

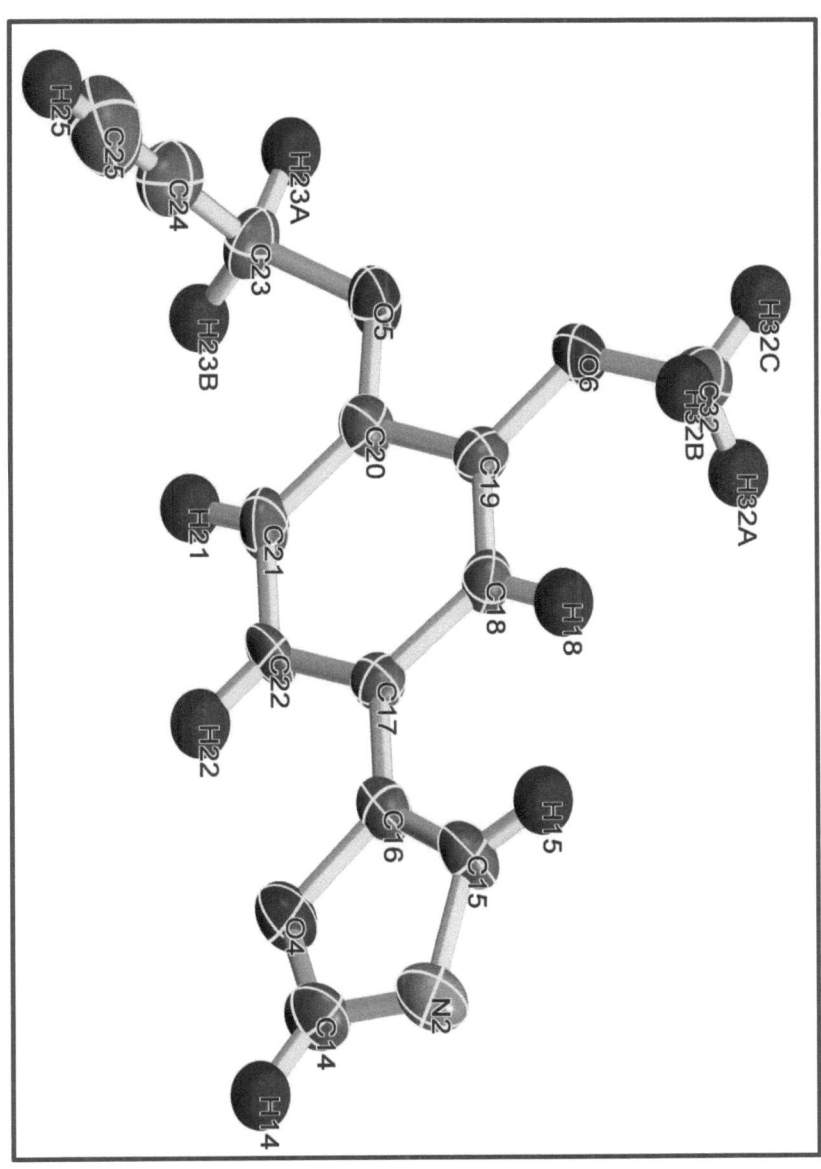

5.3 Packing diagram of the molecules when viewed down the b axis

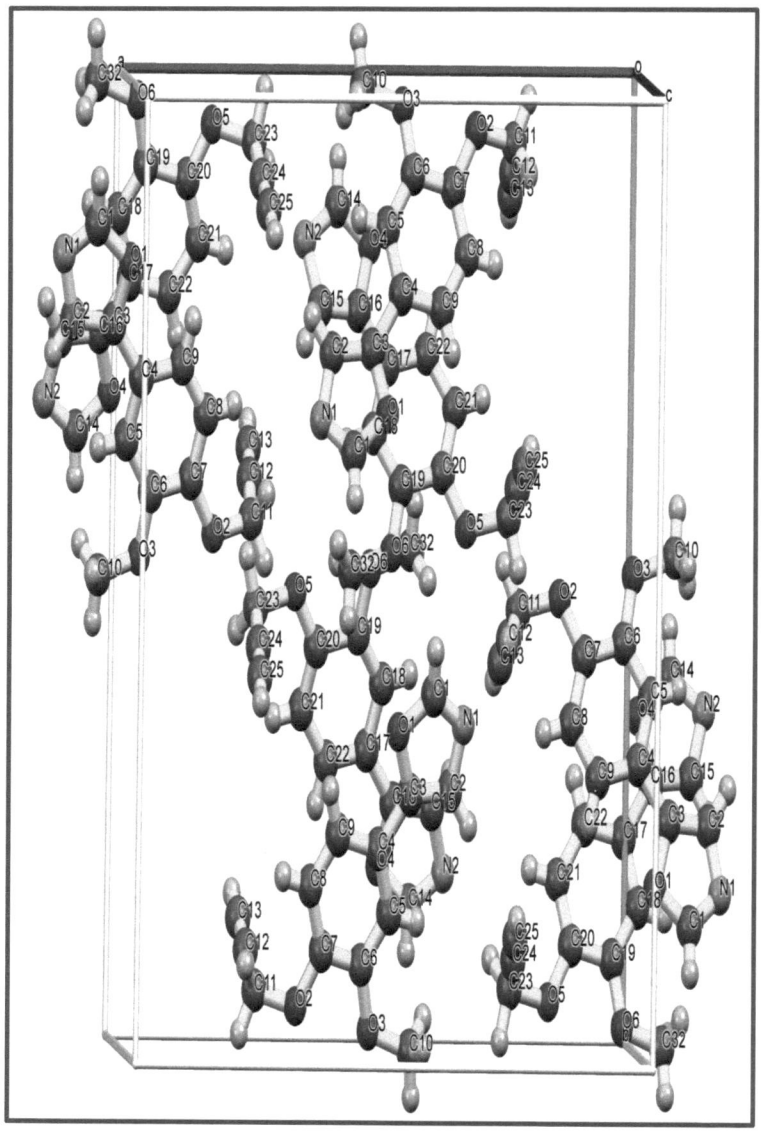

5.4 Packing diagram of the molecules when viewed down the b axis (only one molecule)

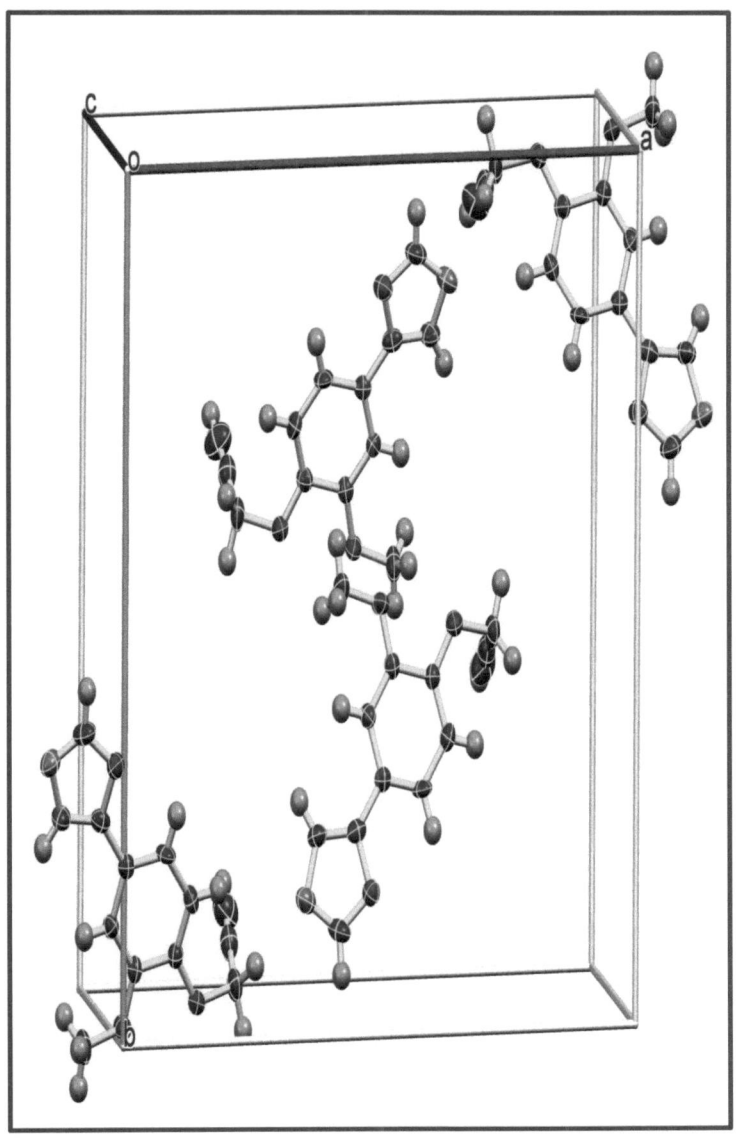

5.5 Packing diagram of the molecules when viewed down the a axis (The dashed lines represent the hydrogen bonds.)

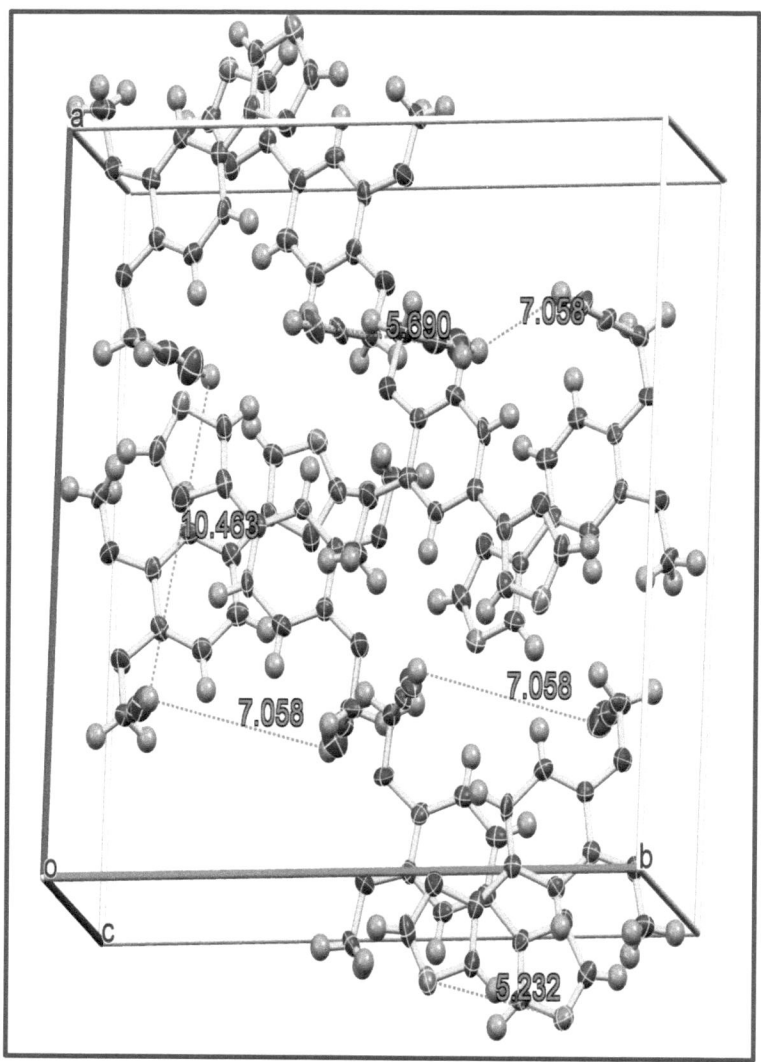

5.6 Packing diagram of the molecules when viewed down the a axis (The dashed lines represent the hydrogen bonds.)

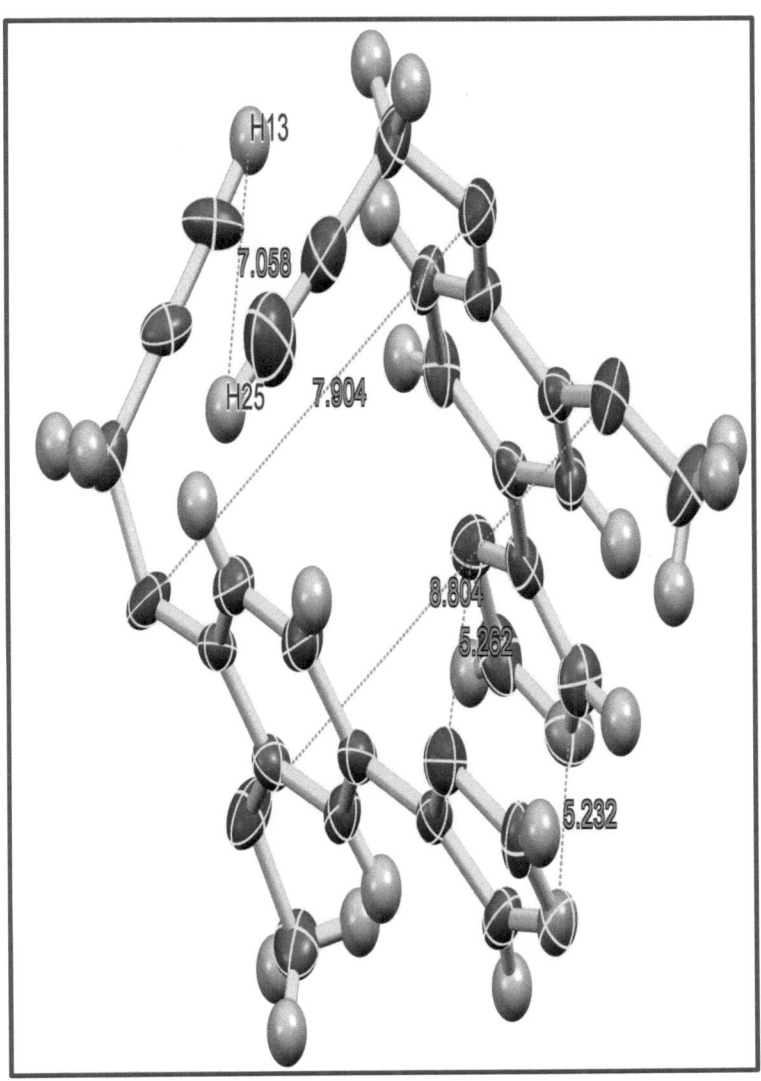

31

6. References

1. Heterocyclic Chemistry TL Gilchrist, The Bath press, 1985.

2. Hartner, **Comprehensive Heterocyclic Chemistry II**, **Oxford**, 1996.

3. Boyd, **Comprehensive Heterocyclic Chemistry II**, **Oxford**, 1984.

4. Ibata, **Bull. Chem. Soc. Jpn**., 1986, 59, 3197–3200.

5. Matsuo, **Chem. Pharm. Bull**., 1972, 20, 669–676.

6. Koyama, Y., J. Agricultural and Biological Chemistry 1981, 45, 1285.

7. Shin, **Journal of Microbiology and Biotechnology**, 2007, 17, 1403.

8. Hanssen, **J. Angewandte Chemie International Edition**, 2012, 51, 12238.

9. Pflugrath, **Acta CrystallogrD**, 1999, 55, 1718

10. Altomare, **J Appl Crystallogr**, 1994, 27, 435

11. **Least Squares function minimized**: (SHELXL97)

12. Standard Deviation of an observation of Unit Weight: $[Sw(F_o^2-F_c^2)^2/(N_o-N_v)]^{1/2}$

13. Cromer, **The Kynoch Press**, 1974, Table 2.2 A

14. Hamilton, **Acta Crystallogr**, 1964, 17, 781

15. McAuley, **Kluwer Academic Publishers**, 1992, Table 4.2.6.8, C:219

16. Creagh, **Kluwer Academic Publishers**, 1992, Table 4.2.4.3, C:200

17. Creagh, **Academic Publishers**, 1992, Table 4.2.4.3, C:200

18. CrystalStructure 4.0, **Rigaku Corporation**, Tokyo, 2000-2010

19. Sheldrick, **Acta Crystallogr A**, 2008, 64, 112

Contributing Authors

The book opens with Articles by Dr. Krunalkumar R. Mehariya, Mrs. Bhagwati K. Gauni Mehariya, Dr. Dipak Vachaani, Mr. Kirtikumar Goswami and Mr. Sunil Galani and Dr. Ranjit S. Pada.

This Book Article will be suitable for Graduate, Postgraduate Chemistry students and other equivalent standard. A Book Article should be plausibly small, but essential aspects of the subject of Practical Physical Chemistry Experiments may not be neglected, usual foundations must be considered, and modern developments should be included. This introductory text is an attempt to present.

The basics of the subject are explained carefully and thoroughly, with standing on type of novel molecules synthesis, characterization and analysis along with reporting of data.

This Book Article would not have been potential without the authors who graciously wrote tremendous chapter for this book. We thank all of them for their outstanding contributions.

Dr. Krunalkumar R. Mehariya

Mrs. Bhagwati K. Gauni Mehariya

Mr. Kirti kumar Goswami

Dr. Dipak Vachhani

Mr. Sunil Galani

Dr. Ranjit S. Pada

About the Authors

Dr. Krunalkumar Ramanlal Mehariya is Assistant Professor at Gujarat Arts & Science College, Ahmedabad since July 2016. He is a basically Organic-Medicinal Chemist and has a vast knowledge of Medicinal Chemistry as well as sophisticated instrumentation.

Mrs. Bhagwati K. Gauni Mehariya comes from the Medical-Microbiology field; she has a great knowledge in Drugs analysis as well as in the Microbiology field.

Dr. Dipak D. Vachhani has great knowledge in Organic Chemistry as well as in Pharmaceutical Instrumentations.

Mr. Kirti kumar Goswami is a basically Chemistry Lecturer and has more than five years of experience in various fields of Chemistry as well as in the Medicinal field.

Mr. Sunil Galani is a good Researcher in Organic chemistry as well as in the Medicinal Organic field.

Dr. Ranjit S. Pada is a good Researcher and is working as a Senior Research Associate at Intas Pharmaceutical Ltd., Ahmedabad, Gujarat.